YOUR KNOWLEDGE HAS VALUE

AF303330

- We will publish your bachelor's and
  master's thesis, essays and papers

- Your own eBook and book -
  sold worldwide in all relevant shops

- Earn money with each sale

Upload your text at www.GRIN.com
and publish for free

Bibliographic information published by the German National Library:

The German National Library lists this publication in the National Bibliography; detailed bibliographic data are available on the Internet at http://dnb.dnb.de .

Imprint:

Copyright © 2014 GRIN Verlag, Open Publishing GmbH
Print and binding: Books on Demand GmbH, Norderstedt Germany
ISBN: 978-3-656-74782-6

This book at GRIN:

http://www.grin.com/en/e-book/280795/hydrograph-recession-analysis-methods-and-its-comparison-using-unsaturated

Rajib Maharjan

Hydrograph recession analysis methods and its comparison using unsaturated moisture movement model

GRIN Publishing

Hydrograph recession analysis methods and its comparison using unsaturated moisture movement model

Rajib Maharjan
Water and Environment,
University of Oulu
Oulu, Finland

ABSTRACT: Groundwater plays an important role in feeding springs and streams, supporting wetlands and land surface stability. In Finland, most water is held in the soil than the surface systems. Hence, Finland's water resources depend on groundwater and biogeochemical processes. The study of groundwater in peatland is important for maintaining ecological balance and conservation of water resources. The groundwater level is one of the key indicators of aquifer conditions and groundwater basins. It helps to interpret hydrogeology, groundwater flow, groundwater sustainability and land usability. The study tries to analyze ground water recharge on peatland catchments using hydrograph recession analysis. The equation for the hydrograph recession curve can be utilized to predict groundwater recharge during each recession period. The steps involved during recession curve analysis includes selection of analytical expression, derivation of recession characteristic and optimization of the parameters. While computing groundwater recharge with recession curve, the high variability of each recession segments creates major problem. Each segment shows the outflow process which creates short-term or seasonal influence. The variation in rate of recession which causes problems for derivation of recession characteristics. The computer software such as hydro-office, VBA macro excel and Matlab are used for recession analysis. The results obtained do not consider climatic influences. The results were then confirmed by using water balance model and statistical tests. The e-water toolkit is used for water balance model and statistical tests are performed using R-software. The rainfall-runoff data are used as input to the software used in each method. From the analysis, required output recession parameters are obtained for further calculation. These estimated recession parameters can be used to predict low flows (groundwater contribution to runoff) to understand catchment groundwater resources and as inputs for the rainfall-runoff model analysis. Hence, the objective of this study is to analyze groundwater recharge by studying the recession limb of the runoff hydrograph. The study work compares various recession analysis methods. It also tries to identify the better method by comparing groundwater recharge from different methods with groundwater recharge from unsaturated water balance model. Furthermore, the recession constants obtained from different methods are compared with the theoretical values. Statistical tests are used to identify the best method among recession analysis methods used in this study.

KEYWORDS: Runoff Hydrograph, recession Parameters, Runoff, Groundwater recharge, unsaturated water balance, Statistical tests

CONTENT

1 Introduction 3

2 Site Description 3

3 Methods 4

   3.1 Hygrograph recession analysis 4

      3.1.1 Individual Recession Analysis 4

      3.1.2 Master Recession Analysis 6

      3.1.3 Wavelet transformation 7

      3.1.4 Recession constant and recharge from baseflow separation 8

      3.1.5 Recession constant and storage from specific yield 8

   3.2 Unsaturated moisture balance components 9

4 Calculations 10

   4.1 Hydrograph recession and Specific approach 10

   4.2 Recharge volume from unsaturated water balance 10

5 Results 11

6 CONCLUSION 13

Acknowledgment 15

References 15

1 INTRODUCTION

Peatlands are major important part of global ecosystem. It shows significant interaction with natural hydrological system, biogeochemical cycling and terrestrial as well as aquatic biodiversity. In Finland, peatlands have high influence in ecological as well as socio-economic aspects. It covers one-third of Finnish land area which is 2.0 million ha of 9.3 million ha [1]. The hydrological study is used to develop the functions and process related peatlands system. Hydrological study is an important part of environmental and ecological study in Finland. In peatlands as in other soil formation there is interactive connection between the surface and subsurface hydrological water system. This study intends to calculate yearly groundwater recharge of two catchments using recession hydrograph. It includes study of various hydrograph recession analysis methods. It also includes various climatic factors that influence runoff hydrograph. The amount of water received by catchment is disintegrated in different time period. The hydrological features of catchment influences runoff and water storage in the catchment. The runoff generated is highly influenced by upslope contributions from surface flow as well as interflow. The study of hydrological behavior in surface and subsurface of two peatlands catchment is the major objective of this study.

2 SITE DESCRIPTION

The two catchments studied in this study are Marjasuo and Röyvänsuo with four years data (2010-2013). Marjasuo peatland has been drained since 1968 for forestry and was restored in 2011.Röyvänsuo is a pristine peatland located in Isosyöte National park. Both of the study catchments are the part of larger Iijoki catchment [2]. The catchments lie in northern Finland at Taivalkoski municipality and both are state owned. The geographical locations of the catchments Marjasuo and Röyvänsuo are at 65o48'19.79'' latitude and 27o48'42.246''longitude and 65o49'12.213'' latitude and 27o48'13.978'' longitude respectively. Marjasuo covers land area of 65ha (0.65km²) and Röyvänsuo 75ha (0.75km²). The two catchments contain almost similar terrestrial and soil formations. Marjasuo has 2.27 ha (3.5%) open water or pond, 30.55 ha (47%) mineral soil, 16.5 ha (25.5%) fen or open mire and 15.6 ha (24%) forested Peatland and paludified forest. Similarly, Röyvänsuo contains 0.5 ha (<1%) open water, 44.25 ha (59%) mineral soil, 18.75 ha (25%) fen (open mire) and 11.25 ha (15%) forested peatland and paludified forest. The map with location of two catchments are shown in Fig. (1)

Fig. 1. Catchment location map [3].

3   METHODS

The recession constants and groundwater recharge of both sites Marjasuo and Röyvänsuo are computed using the recession analysis method and specific yield method. In this study the groundwater recharge obtained from unsaturated water balance model is considered to be precise. So, the groundwater recharge obtained from recession analysis methods and specific yield are compared to groundwater recharge from unsaturated water balance model. Finally, statistical tests are carried out to observe the significance of the results obtained. The basic approaches used for this study are discussed in the following sections.

3.1   HYGROGRAPH RECESSION ANALYSIS

3.1.1   INDIVIDUAL RECESSION ANALYSIS

The analysis of individual recession curves is carried out using RC 4.0 tool from hydro office software [4]. By using RC tool, an individual recession segments are separated from runoff hydrograph. The runoff data is the major input and rainfall is optional. An individual recession curve is selected for short time period with small numbers of declining runoff values. In an individual analysis there is different flow constant for the slow and fast runoff. It consists of two linear models. One represents fast flow and the other represents slow flow. For each model, recession curve is divided into two portions (upper and lower).The initial flow and constant (k) are given by user. For the calibration of individual recession curve, a tool called single calibration in hydro office software is used. The outputs obtained from the software are initial flow, recession coefficient and recession time days. The output obtained is used to calculate final discharge and groundwater recharge during recession period.

i) Final runoff at time t

The output contains two initial flows and two constants for same time period. The initial flows and recession constants are added to get total flow and total constant for each individual recession curve. From total initial flow and total recession constant final discharge is calculated. The flow at the end of recession period is given by equation (1) [5]:

$$Q_t = Q_o K^t \qquad (1)$$

Where Q_t is runoff at the end of recession period   (m3/s) per unit area

Q_o is intial recession flow (m3/s)

t is recession period (d)

ii) Change in groundwater recharge

The runoff and runoff time to complete one log cycle is given by equation (2) and (3) [6]:

$$Q = Q_o 10^{-t/t_1} \qquad (2)$$

$$t_1 = t \ln(10)/\ln(Q_o/Q) \qquad (3)$$

The calculated value t1 is used to find the groundwater recharge between each recession curve. In this method groundwater change is calculated based on each log cycle. Each individual volume is added to get total groundwater recharge volume. Individual ground water recharge volume in each log cycle is the difference between total potential groundwater runoff at beginning of recession and total groundwater potential at the end of recession.

The volume of groundwater runoff at the beginning and end of recession is given by equation (4) and (5) [4]:

$$V_{tp} = (Q_o \times t_1)/2.3 \qquad (4)$$

$$V_r = (Q_o \times t)/(2.3 \times 10^{t/t_1}) \qquad (5)$$

Where Q_0 is runoff when t = 0 (m3/s)

V_{tp} is total potential runoff at beginning (m3)

V_r is total potential runoff volume at end (m3)

t is total time of recession (d)

t_1 is time for 1 complete log cycle (d)

The volume of groundwater recharge is difference between volume at starting and ending of each recession event, which is shown in equation (6) [4]:

$$VR = Vtp - Vr \qquad (6)$$

Where   VR is volume between recession (m3)

V_{tp} is total potential runoff volume at beginning (m3)

V_r is total potential runoff volume at end (m3)

The calculated recharge is converted to daily volume per unit area. The daily recharge volume is calculated as given in equation (7) [4]:

$$V_d = VR/A \qquad (7)$$

Where   A is Area of the catchment (m2)

V_d is storage volume between recession (m per day)

3.1.2 Master Recession Analysis

In master recession curve analysis a single curve is obtained representing all individual recession curves. The analysis of master recession curve is carried out by separating each recession segment from yearly hydrograph. The adapted matching strip method is used for construction of master recession curve. A visual basic spreadsheet macro is used for master recession analysis. In VBA excel sheet, each individual recession curve is fitted to an exponential regression model to draw master recession curve. The date and runoff are initial inputs which gives runoff hydrograph. The automated VBA macro is used to separate individual recession and time of recession from hydrograph. The separation of individual recession is carried out with the separation criteria set by flow duration curve. The flow duration curve shows the percentage of time that a given flow rate is equaled or exceeded. The separation criterion of (10-70) % is selected for each year. The high runoff data value are selected as initial runoff exceeds the corresponding runoff values by (10-70) % in an individual recession curve. The process includes the selection of variable length of recession period from runoff data. The separated time is then ranked in descending order from which initial value of recession is obtained. Then the highest flow value are selected along with declining values. It is then plotted in semi-logarithmic scale in decreasing order which gives the equation with two variables x and y. In the semi-logarithmic plot, x represents time of flow and y represents flow rates. The second highest number gives the second recession. The curve obtained with second highest value is adjusted to the last point value of first recession curve. The adjustment is carried out with segment translation. In segment translation, the time of second recession is shifted to required place along axes till it fit to the end of first recession. The process continues till the last recession curve is combined.

Finally a regression line is drawn with the best fitted model criteria. The regression line obtained is called master recession curve. The criteria are based on trend line R2 describing the data which varies from 0-1. The values approaching to 1 are the best fitted models. The data obtained from VBA macro excel sheet are runoff values for each individual recession, time of recession and equation for regression line. From the exponential regression equation recession constant is calculated. From the recession constant, runoff during each recession and time of recession groundwater recharge is calculated. The calculated processes are similar to individual recession analysis.

3.1.3 WAVELET TRANSFORMATION

Wavelet analysis is carried out to find centred frequency from time series runoff data. Hydrograph interflow is relatively faster than baseflow. The runoff is changed to frequency signal. The time of baseflow is longer. As a consequence, signal frequency is reduced. The central frequency is frequency at which there is change in signal behavior. The wavelength transformation of catchments is done using Matlab. The time series runoff data is converted to frequency signals. To obtain frequency signals from time series data Fast Fourier Transformation (FFT) is used. By using FFT the time domain data is decomposed to frequency signals. From the frequency signals center frequency is obtained. To find center frequency a band pass filter criteria called Nyquist rate is used. The Nyquist frequency and Nyquist rate are two different terms. Nyquist frequency is twice the highest frequency in the signal whereas Nyquist rate is used to obtain symmetric signal. The Nyquist rate is obtained from amplitude modulation which converts signals to symmetric signal within maximum amplitude. In this maximum amplitude is taken as 1. From the symmetric frequency signal center frequency is obtained. The centered frequency obtained from wavelet analysis is used to find the recession parameters for the catchment. The equation for calculation of recession parameter k using the centred frequency is shown in equation (8) [7]:

$$k = e^{-f_c}$$

$$(8)$$

Where   k is recession parameter

f_c is centered frequency from wavelet analysis

The wavelength transformation in this study is only used for comparison of recession constant. The process is relatively new and requires further study to relate with groundwater processes. The further calculation requires initial flow and recession period. The calculation of these flow characteristics further study on reconstruction of original signal. The original signals can be obtained from Short Time Fourier Frequency (STFT) transformation but phase angle cannot be regenerated. Due change in phase angle random data is obtained and data obtained is not equal to original data. The Fast Fourier Transformation (FFT) of signal results in randomization of phase. By doing Inverse Fast Fourier Transformation (IFFT) original signal can be regenerate but at random phase. By using this method the frequency at which baseflow occurs is only obtained. It is unable to determine the original runoff and time at which baseflow starts.

3.1.4 RECESSION CONSTANT AND RECHARGE FROM BASEFLOW SEPARATION

Baseflow is separated from total flow using smoothed minima technique. Baseflow program is used for baseflow separation. A Baseflow program is VBA excel which is used to separate surface and base flow [8]. For separation of base flow mean daily flow is divided into non-overlapping blocks of 5 days. The minima value is calculated for each block. The minima value is called central value. The separation criteria for each bock is as 0.9 × central value <original value. The central value gives ordinate for baseflow line. The process is continued to obtain baseflow ordinates from all values. Base flow index is obtained as the ratio of volume of water lying under base flow line to the volume of water below mean daily flow line (Institute of Hydrology, 1980).

The base from index obtained is used to calculate groundwater recharge volume and recession constant. The equation for calculating groundwater recharge volume from baseflow index is shown in equation (9) [9].

$$BFI = Q_b/Q = R \qquad (9)$$

Where   BFI is Base Flow Index

R is base flow volume (m3)

Q_b = base flow (m3/s)

Q = total flow (m3/s)

The equation for calculating recession constant from baseflow index is shown in equation (10) [10]:

$$BFI = (Ok\,(1-k))/3k \qquad (10)$$

Where   k is recession constant

3.1.5 RECESSION CONSTANT AND STORAGE FROM SPECIFIC YIELD

The specific yield is related to groundwater table. The specific yield and change in groundwater level is used to calculate recession constant and groundwater recharge. The time series graph of groundwater level data is similar to runoff hydrograph. From the groundwater level data, the depletion curves are selected. The change in groundwater level during each depletion curve is used to calculate recession constants. The recession slope is calculated as the ratio of product of specific yield and time to the change in groundwater level as shown in equation (11). From recession slope recession constant is calculated using equation (12) [11].

$$S_y = (\alpha \times t)/\Delta h \qquad (11)$$
$$k = e^{-\alpha} \qquad (12)$$

Where   Sy is Specific yield

t is time (days)

Δh is change in groundwater level

α is recession slope

For yearly groundwater recharge, the average groundwater level change in each depletion curve and average specific yield is used. The recharge calculation is based on the assumption that percolated water immediately goes to storage. This method is applicable for short recession periods [12].The calculation of groundwater volume change is shown in equation (13) [11].

$$G_r = h_{avg} \times S_y \qquad (13)$$

Where G_r = groundwater recharge (m/d)

h_{avg} = Average water level (m)

3.2 UNSATURATED MOISTURE BALANCE COMPONENTS

The water balance model is used for each year. The calculations are carried out with daily data. The water balance model gives yearly groundwater recharge. The outputs obtained from unsaturated moisture balance are rainfall, infiltration, contribution from upslope, total evaporation, recharge and saturated runoff. The groundwater recharge volume obtained as moisture balance output is used to compare the groundwater recharge obtained from different methods used. The outputs from unsaturated balance moisture model are computed using a software toolkit called class U3M-1D. This program uses Richard's equation for water balance calculation. The equation is applicable for any soil, weather conditions or vegetation type. The software toolkit contains three alternatives for soil hydraulic modeling: Van Genuchten soil hydraulic model, Vogel and Cislerova soil hydraulic model and Brooks and Corey soil hydraulic model. Brooks and Corey soil hydraulic model is chosen in this study. Brooks and Corey soil hydraulic model is chosen due to easy mathematical manipulation and flexibility of program allowing user input hydraulic parameters. The outputs obtained from the software are total evaporation (E), saturated runoff (Wdelta), infiltration runoff (Qtop) and infiltration recharge (Qbot). The soil moisture fluxes are separated in upward and downward direction. Total evaporation (E) and saturated runoff (Wdelta) represents upward flow. The total evaporation and saturated runoff lies above arbitrary plane called zero flux plane. The infiltration runoff (Qtop) and infiltration recharge (Qbot) represents downward flow. The infiltration runoff and infiltration recharge lies below zero flux plane.

4 CALCULATIONS

4.1 HYDROGRAPH RECESSION AND SPECIFIC APPROACH

The recession constants and groundwater recharge are calculated from individual recession, master recession and baseflow separation. From wavelet transformation, only recession constants are calculated. The recession constants are compared to theoretical values. The groundwater recharge obtained from various methods is compared to groundwater recharge from unsaturated water balance model. The obtained results are statistically compared. The summary of the results for Marjasuo (Table 1) and Röyvänsuo (Table 2) obtained from hydrograph recession methods and specific yield.

Table 1. Summary of the recharge volume and recession constant calculated from various methods for Marjasuo catchment

Methods	IRS		MRC		Wavelet	Base flow		Specific yield	
Results/Years	Vd (m/d)	k (1/d)	Vd (m/d)	k (1/d)	k (1/d)	R (m/d)	k (1/d)	Gr (m/d)	k (1/d)
2010	0.0243	0.6916	0.023	0.82	0.96	0.0127	0.883	0.022	0.56
2011	0.0495	0.685	0.049	0.86	0.95	0.0206	0.885	0.046	0.62
2012	0.0685	0.6268	0.068	0.51	0.87	0.0273	0.95	0.054	0.66
2013	0.019	0.7766	0.019	0.74	0.97	0.0009	0.996	0.008	0.43

Table 2: Summary of the recharge volume and recession constant calculated from various methods for Röyvänsuo catchment

Methods	IRS		MRC		Wavelet	Base flow		Specific yield	
Results/Years	Vd (m/d)	k (1/d)	Vd (m/d)	k (1/d)	k (1/d)	R (m/d)	k (1/d)	Gr (m/d)	k (1/d)
2010	0.035	0.598	0.038	0.86	0.95	0.0208	0.87	0.03	0.49
2011	0.02	0.834	0.02	0.72	0.91	0.0055	0.94	0.02	0.43
2012	0.0477	0.565	0.047	0.74	0.78	0.0969	0.86	0.03	0.46
2013	0.0213	0.745	0.021	0.76	0.96	0.0132	0.79	0.02	0.66

4.2 RECHARGE VOLUME FROM UNSATURATED WATER BALANCE

The output fluxes are the components of unsaturated water balance model. The output flux contains rainfall (R), total evaporation (E), saturated runoff (Wdelta), infiltration from soil surface (Qtop) and infiltration recharge (Qbot). All the obtained results are in meter per day (m/d). The output fluxes are summed up to get annual values of all components. From results obtained, groundwater recharge volume is separated. The results obtained with all output fluxes in meter per day (m/d) and output soil moisture is shown as Appendix 4. The outputs from unsaturated soil moisture balance are applicable for various hydrological processes. The components from unsaturated water balance are considered to be precise with some uncertainty. So, the recharge volume from soil moisture water balance is compared with groundwater recharge obtained from other methods. From yearly water balance, components for each year (Table 3) and (Table 4) recharge volume is used for comparison. The negative sign in Tables (3) and (4) indicates downward movement of soil moisture.

Year	Evaporation(E)	Sat. runoff (Wdelta)	Recharge (Qbot)	Infil. Runoff (Qtop)
2010	0.000022	0.332178	-0.023804	-0.348257
2011	0.000026	0.497769	-0.049572	-0.403306
2012	0.000023	0.374437	-0.067326	-0.351485
2013	0.000024	0.367651	-0.019379	-0.218359

Table 4: Soil moisture balance components for Röyvänsuo catchment

Year	Evaporation(E)	Sat. runoff (Wdelta)	Recharge(Qbot)	Infil. Runoff (Qtop)
2010	1.73E-05	0.37532	-0.0366	-0.35964
2011	2.11E-05	0.41336	-0.01964	-0.39722
2012	1.67E-05	0.22122	-0.04778	-0.51866
2013	1.68E-05	0.34768	-0.0234	-0.22061

5 RESULTS

Analyzing the results obtained from hydrograph recession analysis and specific yield, the good method is identified. The most efficient method is determined from the following two approaches: a) box plots of recession parameters and recharge volume b) comparison with theoretical recession parameters.

The box plots for recession parameters (Fig. 2 and Fig. 3) show the plot of recession parameters from different methods for two catchments. The plot obtained is compared to theoretical value that ranges from 0.85 to 0.99 [13]. The box plots for (Fig. 4 and Fig. 5) reveal groundwater recharge from different methods and groundwater recharge from unsaturated water balance.

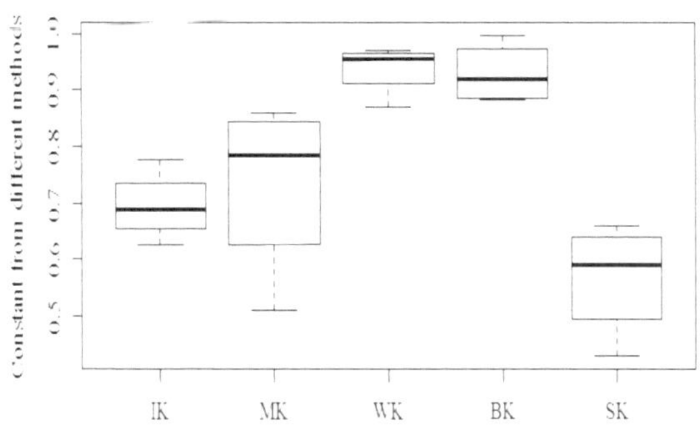

Note: In Fig. (2): BK = Base flow recession constant, IK = Individual recession constant, MK = Master recession constant, WK = Wavelet constant and SK = Specific yield constant.

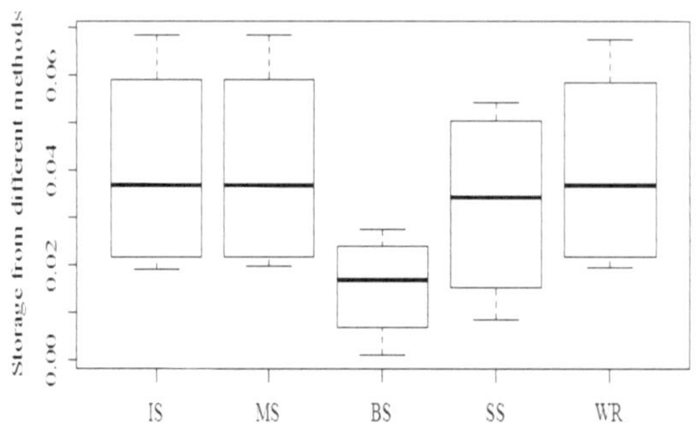

Fig. 3.   Box plot for Groundwater recharge for Marjasuo catchment.

Note: In Fig. (3): IR = Recharge volume from Individual recession, MR = Recharge volume from Master recession, BR = Recharge volume from Base flow, SS = Recharge volume from specific yield and WR = Recharge volume from water balance.

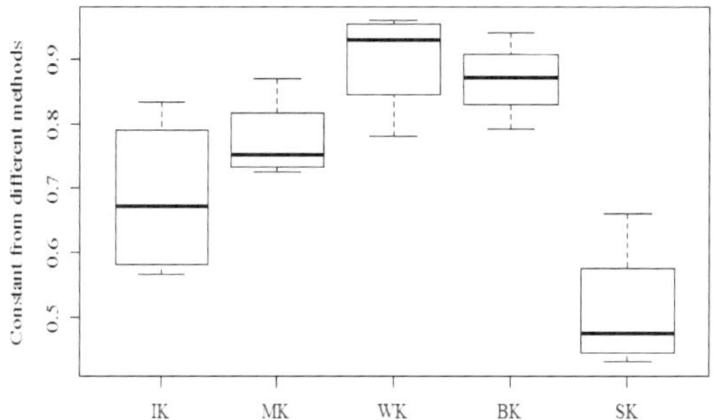

Fig.4. Box plot for Groundwater recharge for catchment.

Note: In Fig. (5): IK = Individual recession constant, MK = Master recession constant WK = Wavelet constant, BK = Base flow recession constant and SK = recession constant from specific yield.)

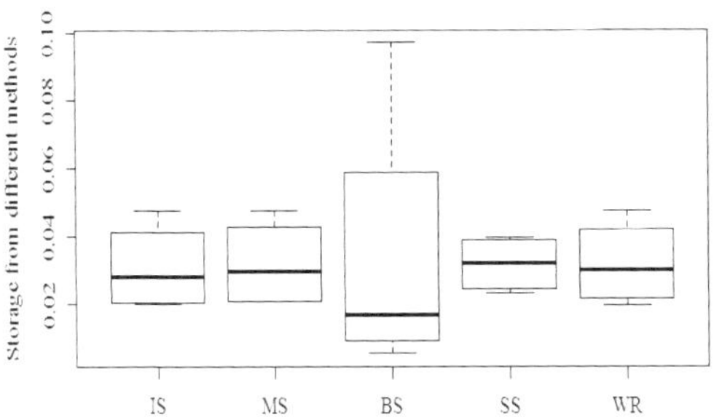

Fig. 5. Box plot for Groundwater recharge for Röyvänsuo catchment.

Note: In Fig. (4): IS = Recharge volume from Individual recession, MS = Recharge volume from Master recession, BS = Recharge volume from Base flow, BS = Recharge volume from specific yield method and WR = Recharge volume from water balance.

The theoretical value for groundwater recession lies in the range of 0.85 to 0.99 [13]. The box plot of recession parameters for Marjasuo (Fig. 16) and 63box-plot of recession parameters for Röyvänsuo (Fig.18) show the calculated results of recession constants. Both catchments show that the parameters calculated using wavelet transformation and baseflow lies within the range of theoretical values. The box plot of recharge volume for Marjasuo (Fig. 18) and box-plot of recharge for Röyvänsuo (Fig. 20) show the calculated groundwater recharge. Both catchments show that groundwater recharge volume calculated from Individual recession and Master recession are close to the recharge values calculated from water balance method. The recharge volume from baseflow method is not close to recharge calculated from Water balance. The recession parameters calculated from Master recession nearly lies in range of theoretical value. Also recharge volume from Master recession is close to recharge calculated from Water balance.

6   CONCLUSION

The study is based on comparison of different methods for calculation of recession constant and groundwater recharge using hydrograph recession analysis. Hydrograph recession analysis of catchments Marjasuo and Röyvänsuo is carried out with runoff data. Different recession analysis methods: individual recession, master recession, wavelet transformation and baseflow separation are used to compute recession constant and groundwater recharge. Wavelet transformation is only used for calculating recession constant. For further implication of wavelet analysis further work is required. The groundwater level change during recession is related to groundwater recharge. The recession constant and groundwater recharge can also be calculated with specific yield and groundwater level. So, the results from specific yield are used for comparison. The water balance parameters are computed by unsaturated moisture movement model. The unsaturated moisture model includes catchments parameters related to the runoff process. The water balance components obtained from water balance models are almost accurate and are used in various land use practices and effective soil-water conservation. The water balance model provides parameters for the purpose of rainfall designs, storage yield, and prediction of meteorological, study of hydrological and ecological processes. The statistical comparison of groundwater recharge method and groundwater recharge from unsaturated water balance model shows master recession analysis method is the most efficient hydrograph recession analysis techniques among the methods used in this study. Hydrograph recession also correlates climatic and geomorphologic features of the catchment. Hydrograph recession parameters contain embedded information about the flow regime and hydro geological characteristic of the catchment. Hydrograph recession analysis should be carried in fast and objective manner. The different recession analysis method used in this study gives recession parameters and groundwater recharge for the catchments. The recession parameters denote the slope of depletion curve. It also represents the rate by which water flow from the catchment to the runoff point. The flow in the catchment is also influenced by catchment slope and climate. The climatic information in hydrograph is further clarified by unsaturated moisture movement model. In unsaturated moisture model, soil moisture content in various soil layers are 67calculated at certain time period. It also contains the direction of flow at certain time. The amount and direction of flow is influenced by climatic condition and hydraulic conductivity of the catchment. The recharge component obtained from unsaturated moisture model is used to compare recharge calculated from different methods.

The recession parameters calculated from various methods differs from each other. The recession parameter depends on the selection of recession curve and procedure of its analysis. The calculation in this study also shows different recession parameters and groundwater recharge values. To identify the best method, statistical tests are carried out. The wavelet transformation is a recent method applied for qualitative analysis of data which gives recession constant close to theoretical values. The study provides adequate information about various methods of hydrograph recession analysis and specific yield by which recession constant and groundwater recharge are calculated.

ACKNOWLEDGMENT

This work is funded by university of Oulu (Water Resource and Environmental Laboratory) and MVTT (Maa-ja Vesitekniikan Tuki). I want to express my gratitude to university of Oulu and MVTT for generous financial support. I will be forever grateful to Professor Björn Klöve (Director, Water Resource and Environmental Engineering Laboratory, University of Oulu), Anna-Kaisa Ronkanen, and Meseret Menberu for their continues help and support.

REFERENCES

[1] Kimmo Virtanen and Samu Valpola. (2011). Energy Potential of Finnish Peatlands. *Geological Survey of Finland*. Special Paper (49), 153–161.

[2] Anna-Kaisa Ronkanen, Hannu Marttila, Juha Siekkinen and Bjørn Kløve. (2010). 'Stable isotopes studies to increase knowledge from the role of peatlands in catchment hydrology', *paper presented to Abstracts and Programme of the Cost Action FP0601 Forman Workshop at the Finnish Environment Institute, Helsinki and Hyytiälä Forestry Field Station, Finland* 6.–8.9.2010, viewed: 18th April2014,<http://www.metla.fi/julkaisut/workingpapers/2010/mwp168.pdf>.

[3] Data SIO, NOAA, U.S. Navy, NGA, GEBCO. (2007). *Google Maps [online]*. [Accessed 18 April 2014]. Available at: <http://maps.google.fi >.

[4] Hydro-Office. (2010). *Software for Water Science*. [ONLINE]Available at: http://hydrooffice.org/Downloads/List.aspx?section=Manuals. [Accessed 07 May 14].

[5] Tallaksen, L.M. (1995) A review of baseflow recession analysis. *J. Hydrol*. 165, 349-370.

[6] Meyboom P. (1961): Estimating Ground-Water Recharge from Stream Hydrographs. *J. Geophys*. Res. 66(4), 1203–1214.

[7] Joko Sujono, Shiomi Shikasho and Kazuaki Hiramatsu. (2004). A comparison of techniques for hydrograph recession analysis. *Hydrological processes*. 18 (3), 403–413.

[8] Morawietz, M. (2007). *Baseflow. [ONLINE]*. Available at: http://www.fi.muni.cz/~xforejt/baseflow/. [Accessed 28 July 14].

[9] Jozsef Szilagyi. (1999).On the use of semi-logarithmic plots for baseflow separation. *Ground water*. 37 (5), 660-662.

[10] Jozsef Szilagyi, F.Edwin Harvey and Jerry F. Ayers. (2003). Regional Estimation of Base Recharge to Ground Water Using Water Balance and a Base-Flow Index. *Groundwater*. 4 (4), 504-513.

[11] Sujay Raghavendran (2013) Estimation of Changes in Groundwater Storage using Recession Curve Method, Karnataka, Surathkal: *National Institute of Technology*.

[12] Crosbie, R. S., Binning, P. and Kalma, J. D. (2005). A time series approach to inferring Ground water recharge using the water table fluctuation method. *Water Resources*. Res., 41, W01008, doi: 10.1029/2004WR003077.

[13] K. Subramanya (2008). Engineering hydrology. West patel nagar, New Dehli: Tata McGraw-Hill publishing. 199-201.